My Life as Seen from Orbit

PART ONE: ORBITS 1-6
1944-1950

Frank White

My Life as Seen from Orbit
Part One: Orbits 1-6, 1944-1950

First printed edition published in 2020 by Multiverse Publishing, LLC.

Print ISBN-13 978-1-960119-83-4
eBook ISBN-13 978-1-960119-82-7

Printed in the United States of America

MULTIVERSE
PUBLISHING
multiversepublishingllc.com

INTRODUCTION

I'm trying something a bit different here, so let me explain the reasoning for it before you begin reading. I'm fascinated with memory and how it works (or doesn't). Over the years, I've come to understand that my memories are to some extent *constructs*. They are neither true nor false, and above all else, they are not factual. For example, for a long time, I had established in my mind that my grandfather had died before I was born. Later, I came to realize that he had died when I was three years old. There's a story about that, and it's in here, but it can wait for now. Anyway, that memory was, for a long time, plain wrong, and I only straightened it out when I went into therapy.

Getting it straight was important, because it led to critical new understandings about how his death had affected me, which pointed to some opportunities for closure. I will have more to say on all of this later.

Moreover, this book has been influenced by several other writers whom I admire.

For example, I recently listened to a book on tape by Merilu Henner. She is one of the few people in the world who have an "extreme autobiographical memory." She remembers, accurately, everything that has ever happened to her. Give her a date and she will tell you what she was doing then. "March 15, 2000..."

"Oh, yes, I was on my way to Italy on a Virgin Atlantic flight. I was in seat 2A."

It is not at all clear why only a few people in the world have this capacity, or if it conveys some evolutionary advantage to those who do. I find this to be a fascinating question that demands answers. Wouldn't it be helpful if we could remember everything? Why do we remember some things that happened years ago so clearly, but do not remember what took place only yesterday? I have also read that memories that cause a jolt of adrenalin to flow through your body are the ones you retain. The adrenalin is a marker that tells you, "This is important, don't forget it." This theory explains why we remember traumatic events, of course, but it does not explain why I recall certain seemingly ordinary things from the time I was four, and the rest of that period is blank.

Anyway, in Marilu's book, she teaches you some of the techniques that will help you not only to have a better memory but also how to use your memory more effectively, which is really the point. She notes that we draw conclusions from our memories that affect our current and future actions, and we can become more effective in life if we learn how to make the right assumptions drawn from past events. That is what really interests me.

Another author of note is Dr. Kevin Leman, who wrote *What Your Childhood Memories Say About You*. This is a revelatory book. It answered a question I had long asked myself: why do I remember one or two incidents from the age of three, and perhaps three or four from the age of four, and nothing else about that period? He says that you recall those incidents that helped you to know who you are and where you fit within the world. That made a lot of sense to me, and explained why certain memories are deeply embedded in my psyche, while there are also years in the past that contain no information at all.

Finally, I have been influenced by *From Age-ing to Sage-ing*, a brilliant book by Rabbi Zalman Schachter-Shalomi. He argues, quite convincingly, that we need to stop being "elderly" and start being "elders," taking our rightful place as thoughtful members of the modern "tribe." He considers writing one's biography to be valuable as a way of gaining perspective on the wholeness of that life, as well as contributing one's wisdom to future generations.

In writing about my past, I have found that the process triggers a more complete set of memories and insights about what was going on at the time. I have also seen that Merilu Henner is right. I have drawn conclusions at the time that affected my behavior later, but with hindsight I can now come up with more meaningful insights into what went on then, and how it affects me now. In other words, as Merilu points out, if you become conscious of the conclusions drawn from your memories, you can reevaluate and change them. You can be a creator of events, rather than a victim.

Another influence on this book is my own work on the Overview Effect. This is the experience of astronauts and cosmonauts as they see the Earth from space and in space, from orbit or the moon. A number of my colleagues, including Paul Mayewski, Kevin Kelley, and Alex Howerton, have pointed out that there is not only an overview of space but also of time. As I moved deeper into this writing, I began to see my life as if I were in orbit, looking down on it, seeing it as a whole, and understanding the connections that existed within it. In short, I began to perceive the "Overview Effect of time."

This insight led to the title of the book, *My Life as Seen from*

Orbit. Please bear in mind that this is not a complete autobiography. While I have been alive, the Earth has gone around the sun 70 times. As measured in conventional terms, that is almost 595,000 hours of experience. Since I do not have extreme autobiographical memory, I am probably going to be reporting on less than five percent of the total.

I also came to realize how short our lives are. When I was born, the Earth had gone around the sun 4.5 billion times. By the time my body dies, my 70 plus years on Earth will hardly register in cosmic time. Anyone who is asked the age of our planet will still say 4.5 billion years because adding in another 70, 80, or even 200 would make relatively little difference. Short though they may be, of course, our lives are truly ours, are therefore important to us, whether the universe notices or not.

Those of us who are of a religious or spiritual leaning believe that there is a God or Higher Power who does care about each and every one of us, and that means something as well. I believe that, and I also believe that my life, though short, makes a small contribution to the whole, which is of God's making, and therefore important.

So, going on the assumption that my life has meaning, I am taking the time to document what I have learned while living it. Moreover, if my life is truly about making a contribution to the whole, this research should help me to make a more meaningful contribution.

This is, then, a collection of important memories that I have retained during my short time as a member of the crew of "Spaceship Earth," a brilliant metaphor coined by one of my heroes, Buckminster Fuller. I am looking for events that stand out, and I'm assuming that they stand out because I drew conclusions from them. At the end of each of the longer section, I am going to see what assumptions I made based on the experience, examine those assumptions to see if they might be altered, and then consider how changing those assumptions might change my life for the better going forward. In a very real way, this act of looking into the past is a tool for shaping the future.

Part One: Orbits 1-6
(aka 1944-1950)

CHAPTER ONE
BIRTH

I was there, but of course, I don't remember it directly, since I was just being born. My mother told me about it, though, and here's how I've always imagined it:

It was a hot spring night in the Mississippi Delta, the third year of the war, 1944, and the men were all off in Europe or the Pacific, fighting the threat of Nazism, Fascism, and Japanese Imperialism. The women had to cope as best they could, usually alone. On this night, my mother, a beautiful young woman with flowing red hair, lay crying on the operating table in the hospital. She was ready to deliver her son, me, her first and only child, but the doctor wasn't there. These were not the enlightened days of home births, water births, midwives, and Lamaze. These were the days when the doctors ruled, and what they said and did went unchallenged. (Of course, the doctors were all men.)

"Why are you holding my legs together," asked my mother. "I'm pushing, I'm ready!"

"Just hang on," said a near-frantic nurse, "the doctor isn't here yet." "Well, where is he?"

"I'm sorry, I don't know."

The doctor was finishing his dinner and had told the nurses to prevent my birth from taking place until he arrived. The pain experienced by the mother-to-be didn't matter to the doctor at the time. He was hungry and he wanted his dinner. There weren't many physicians around in those days, and he was old now. He had delivered a lot of babies, and one more coming a little bit later into the world wouldn't make much difference anyway. His daddy was off in the war, and the poor child would most likely be an orphan anyway. Lots of the men were getting killed in the Pacific and in Europe. Might as well finish his dinner.

The doctor finally arrived at the hospital, and I emerged into the world at 8:30 pm on April 3, 1944. My father was, indeed, not there, but my mother was (of course!), and my grandmother, and my beloved grandfather. My aunt, who would deliver my cousin 27 days later, waited for them at home. It was not an ideal way to begin my life, but it could have been worse, and it was the way of lot of kids started out in those days.

Looking back on it, I'm aware that a lot of babies were born without fathers during those years, and not just in the United States. So my beginning was probably repeated, with variations, worldwide, and a lot of mothers had to give birth without any doctors or nurses being there at all. I wonder how that has shaped the world we live in today.

CHAPTER TWO
GROWING UP SOUTHERN

Because my father was away, fighting in the war, I grew up in a kind of commune, though no one would have called it that at the time: mother, aunt, grandmother, grandfather, first cousin, and me, all in a big house on Strong Avenue in Greenwood, Mississippi. Later in life, I couldn't remember much of that time in detail, though I have always thought of it as a happy interlude. What I did remember was my father coming home. That should have been a pleasant time as well, but it really wasn't.

I do have a memory of being a baby and lying on my back with my first cousin, Anne, in the back yard of the house. Later, I found a picture that showed it! I have to admit I can't prove that the memory didn't come from the picture, rather than vice versa, but I really believe the memory came first. As I have said, though, memory is so strange.

(Anne is now known as Harriet, her first name. Like me, she was called by her middle name for years until she finally started using her first name in her adult life.)

Anyway, it wasn't all that easy when my father did return. Contrary to what the doctor might have imagined, he didn't get killed, although it was a close call when his troop transport was torpedoed. He survived a couple of days in a lifeboat on the open ocean, but sustained an injury to his hip that would remain with him for the rest of his life.

Of course, I didn't know any of that at the time. All I knew was that, suddenly, I didn't have my mother to myself any more. She seemed distracted by this *stranger*, she seemed to want to please the new guy more than me. As a little kid of maybe 1½ years old, I had never known a father, except for my grandfather, and now, this unknown person came into our lives and changed *everything*. First of all, we moved to another part of town, living with some friends while our own house was being built. That was kind of like the commune, but not exactly. My parents' friends were nice, and the husband, who was fat and sat around in his undershirt a lot, was also kind of funny. I remembered that whenever this guy, whose name was Joe, wanted me to be quiet, he would say, "Pipe down, boy," and laugh. He sounded angry, but he really wasn't. I liked the attention, but I didn't like being told to be quiet.

That little house was crowded, and I knew, even at a young age, that we were imposing on my parents' friends, and that everyone really looked forward to the day when our own home, right next door, was built. The new house went up without too much trouble, except one day, my mom and I were taking a nap, and when I woke up, I could hear her shouting at someone next door. I went out to see what was going on, and found my usually quiet mother screaming at this painter person.

"That's the wrong color, it's all supposed to be white! The whole house is supposed to be white!"

I was too young to get it. I thought the brown color looked nice on the house. It was just one side, after all. But my mother was very angry, and very sad at the same time. She felt that the painter didn't care, and had just ignored her. The painter, a skinny little fellow in paint-splattered overalls, didn't seem perturbed or apologetic, and it was clear he didn't plan to do it over. Well, this was 1947, and she was just a woman, after all. I didn't know anything about it, but that's how men treated women in those days.

It still seems remarkable how many memories I have retained of that little house in Greenwood, Mississippi. Even when I can't remember things that happened last week I can recall some of the events of that long-ago time.

For example, I remember that it snowed once, and everyone was completely blown away by it. They rarely saw snow in those parts, and no one knew quite what to do. I recall that my father and some of his friends pushed a car out of a ditch while I watched. The men had trouble standing up because they didn't have the right kind of shoes for the weather, and the car's wheels spun because they had no special treads for winter driving. Again, I later found pictures of that winter, confirming the surprising snowfall.

I also remember the day I thought I was going to die. Out playing next to the wrongly painted house, I suddenly felt horrible. I was very young, with little experience of illness, so this felt very scary to me. Trailing dread behind, I entered the house (I thought for the last time) and told my mother, "I'm dying."

"What?" she said, looking at me with concern.

"I'm dying, I've never felt so bad."

My mother felt my forehead and took a closer look.

"You're not dying, honey, you have the flu."

Of course, I had never heard of the flu, but felt relieved to know I wasn't dying. Then, I went to bed for a long time, and my mother took good care of me. It was awful, but at least I got to drink Coca-Colas more often than usual. That, at least, was a side benefit of the flu. Later in life, I would learn that it was called a "secondary gain" of illness. For the rest of my life, whenever I have felt sick, I've had an irresistible urge to drink Cokes, even though they are no longer on my list of foods/drinks to consume on a daily basis.

CHAPTER THREE
BETRAYAL

There came the day when my father didn't go to work. That was odd. Dad was there in the bedroom, putting on a sleeveless undershirt. I remember it very clearly.

"Is something wrong, Daddy?"

"No, we're just going somewhere special."

"Where?"

"Don't worry, you'll see."

The next thing I knew, we were in a very strange place, huge and alien. It was special, all right, but not in a good way. We stood before a desk that stood in turn before an endless-looking hall, dark and foreboding. A woman dressed all in white asked my mother and father questions. They seemed calm, but I was terrified. Then another woman, also in white, came to take me away down the long hall, and I screamed as I saw my parents' figures receding in the darkness. They came rushing to me when I shouted, and told me it would be all right, but it wasn't, of course. I didn't know it then, but it was the same hospital where I had been born.

To this day, years and years later, I have other memories of that horrible time. I recall being strapped down and wheeled down the long corridors again by the women in white, the nurses. I screamed, I cried, but no one came to help me. Where were my parents? Why didn't they save me? Then came the moment that I was taken into the operating room. I had my chance when they moved me to another table and undid the straps. One kick, two, a connection, and *wham*, the tormenting nurse jumped back. "Doctor, do something!"

And then blackness, but satisfaction. I had gone down fighting. I'll never forget how good it felt to punish my tormenter and to hear her cry out to the doctor for help. To me, she was just another bully who'd gotten what she deserved.

I woke up back in the room. My throat burned, but now people were nice to me. My mother fed me ice cream and told me how sorry she was about what I had endured. She even let me get out of bed, but the mean woman in white appeared at the door, as if she had been waiting to catch me at something, and said, "You shouldn't be out of bed. Get back in there, right now." So much for feeling that everything was going to be all right! I could see that as long as that nurse was around, I had to be careful. I started to learn, at an early age, to be cunning, especially when in a hospital. I would use this insight, many years later, to support Donna when she had to stay in the hospital for long periods.

What had happened to me was commonplace at the time: I had had a tonsillectomy. Back then, when a kid's tonsils became swollen and inflamed, the standard procedure was "Yank 'em out."

CHAPTER FOUR
ABANDONMENT

Eventually, the horror ended and I was back home, but I had to stay on the couch and get well slowly. I heard that my grandfather was coming to see me, and my heart leapt. He always made things right, didn't he? Outside, they were paving the streets, and the pungent, unpleasant smell of blacktop permeated the small house. My grandfather did come, but he did not stay long, nor did he make things better. He just wasn't himself. He seemed distracted, remarkably uninterested in my trauma. What was wrong? First, my parents had failed me, now my grandfather? I did not know it then, could not know it then, but my beloved grandfather would be dead within the week. He took his own life.

Looking back on that time, during therapy, I realized that my grandfather's suicide represented the ultimate trauma of my early years. In talking about the event with the therapist, I understood, for the first time, that I had lied to myself about his death, creating the fiction that it had happened before I was born. I had entered into a kind of amnesia after his passing, then "woke up" in kindergarten. I had gone through about a year that I no longer recalled, but in my later years, I believed the fabrication, even though I simultaneously knew the truth. I guess it was the ultimate in "compartmentalization."

I felt that my grandfather had abandoned me, and I was deeply hurt by that. But I suppressed it, pushed it down so far that I couldn't access it, except later on, when others left me, or rejected me. Then I had the same feelings of being unworthy, helpless, and confused. This was especially true if someone that I knew took their own life, which has happened more than once. Abandonment became one of my greatest fears.

I still have that fear. It continues to happen, of course, though I have a little bit more maturity about it than I did as a child. Now, I experience the same feelings of panic and powerlessness, but I understand that these things happen, and it's not all about me. It is about the other person as well. I have gone through life with the theory that if I am a nice person and do the right thing with others, they won't betray me or abandon me, but the theory is proved wrong again and again. Old habits are hard to break, though, so I cling to the notion that I haven't been nice enough or done enough of the right things.

As I am writing this, I have come to realize that the fear may never go away, and that I may never understand why people I consider to be close friends drop me or ignore my phone calls and emails. I suppose I will never get past that feeling of stress and pain in my gut when it happens, but at least I might have more awareness of why it is happening to me.

Anyway, I remembered my mother saying everyone loved my grandfather, but that he was just too nice, too good. I found out, too, that he had left a note. Many years later, my cousin Anne, who had lived with me in the "collective," told me the note said something like, "I'm so tired. I never got over the war. I'm sorry. I'm going home now." He had been in World War I, but I knew little about it. Later, it seemed my father would never get over World War II, and I myself would never get over Vietnam.

The price to be paid for being an American man had to be paid again and again, in every generation. In our time, women are fully integrated into the armed forces, and I suppose they are not getting over it, either. It seems we just don't want to recognize the human cost of war, on any level. If we did, and if we ended the collective delusion that war is glorious, I suppose wars would stop. I think it is a really good idea to put people who have been in combat in charge of a country's policies. They seem to be the ones least likely to send others into wars. I also think that we ought to have the draft, because with a professional army, we simply pay for others to experience the pain and trauma. We are insulated from it.

Over time, and through therapy, I have begun to access my love for my grandfather, which still lay there, under the trauma of his death. As I've held onto that, I've healed, just a bit. I have finally put his picture up in my study, and I can look at it with more love than pain. I think I've come to understand that what he did was not connected with me, not really. While I don't know all of the story, I've become a lot more forgiving of *everyone's* human weaknesses, my own included. Anne tells me he used to play with us and we loved him a lot, back on Strong Avenue in Greenwood. She also reminded me that his car dealership burned to the ground, and he had no insurance for it, so that ruined him financially. I have a memory of him standing at the edge of the lot, crying. I honestly don't know if it is real, but I have it.

Anne says we have to let people go when they make choices like that, and I agree. I know now that he didn't mean to hurt me. However, he was hurting so much that he saw death as his only choice in life. As I have grown older, I can understand that better.

Now, I am trying to experience my love for him and also the happiness I had with him for three years of my early life. In terms of the present, I am looking to avoid having every experience with other human beings be influenced by whether they embrace or abandon me. It is not easy, though, to let go of the fundamental way in which I have approached the world since the age of three.

I need to sort out how much of it is really me and how much of it is a reaction to the trauma I suffered at the age of three. Dealing with this realization will be a lifelong effort, I suspect.

CHAPTER FIVE
THE LONER

In contrast to my earlier life, I remember *a lot* about kindergarten. The main recollection focuses on being an outsider, as opposed to how I felt back in my neighborhood. The boys thought they were cowboys, and they ran around together in little posses. My friend, Ralph, who lived next door, and whom my cousin Anne called "Goddy Ralph" for some strange reason, thought of himself as an Indian, and the cowboys gave him a hard time. I found that I could not come up with a description of myself. I knew I felt a lot of sadness, but since I had suppressed the death of my grandfather, I did not know why. I was not a cowboy, nor an Indian. I was just myself, but what did that mean? It meant I was conscious of myself, while I didn't think the others were! I am embarrassed to say I thought of them as inferior to me, but much happier. I regret saying this, but I sometimes still have those feelings as an adult. It's become a paradigm, or filter, through which I see the world.

I may be totally wrong, and remember, this is a perspective created by a little kid, but I tend to believe that other people are not as thoughtful as I am, not as inner-directed, but they are happier for being that way. Of course, since I don't have ESP and cannot read minds, I don't know that this is so. I have no idea what other people are actually thinking, and I can't even prove that they are.

Anyway, on one occasion, I stood next to the flagpole, just thinking and wondering. Time kind of stood still, and I asked myself, "What am I doing here? What is my life's purpose" I experienced, even at so young an age, a sense of self as a finite being in a world I did not understand. No answer came, so I went home and asked my mother if she knew the meaning of life. She said, "I don't know, honey." She was simply being honest, but it was a great disappointment to me. If the adults didn't know, who did, and how did you find out what the meaning of life might be?

Like many children of that age, I had imaginary friends. Their names were "Nobody" and "Nothing." I couldn't see them and they had no form, but they were presences and they were comforting. I recall that they were different from one another, with Nothing being less substantive, and Nobody being a little heavier. I thought of Nothing as being more intelligent than Nobody, but a little less friendly.

In my mind, I was quite clever to name my imaginary friends Nobody and Nothing, because it implied I knew they weren't real. This was the case as opposed to calling them "Jack" or "Tommy" or some kind of real name. I would occasionally tell people about them and wait to see if they go the joke, but none of them did. That was another great disappointment to me.

Perhaps the most memorable day in my kindergarten career came in year two when the teacher announced who would be the valedictorian that year. I knew all about that, because I remembered it from the previous year. Everyone in the class had a workbook that they used for learning each day, and at the end of the year, the child with the best one was selected to be valedictorian. In later years, when I told the story, almost everyone said, "You had a valedictorian for kindergarten?"

They were shocked that a little kid would be put through something like that.

"Yes, we did," I would say.

On that particular day, she stood in front of the room and announced that they had selected the valedictorian. Instantly, I *knew* I was the one, and also knew that I did not want to be it.

When she announced that I was indeed valedictorian, I immediately said, "I won't do it," but she just laughed and said, "Of course you will." That was merely the beginning. I kept on insisting that I wouldn't give the speech, while also working on memorizing it (I didn't have to write the speech; the teacher took care of that. I just had to deliver it.) My parents and relatives begged me to go ahead and do it. One of my kinfolk promised me a silver dollar if I gave the speech. That was my first bribe, and my first compromise to please others.

I am sure that the speech was not that long, but to me, at the time, it probably seemed endless and impossible to remember in front of a bunch of people.

Looking back on it, this was clearly one of the most formative moments of my life. I was a shy kid who liked nothing more than to be left alone. I was still trying to deal with the arrival of my father and departure of my grandfather. All of a sudden, because of certain talents I had, the world was thrusting me into the spotlight. *And no one seemed to care how I felt about it.* "This is a big honor, Frank, just get on with it!"

Anyway, on the big day, I did deliver the speech. My main worry was that my mortarboard would fall off when I took my bow at the end. The teacher insisted that I had to take that bow, but she never told me how to keep the "hat" on. So I had to improvise, folding my right hand over my belly, quickly dipping forward while holding the thing on with my left. I was angry that no one had told me how to do it properly. Graduating and being valedictorian in kindergarten revealed the two essentially opposed elements of my nature: on the one hand, I liked the limelight, and in later years, I found that I possessed a special ability to move people through my public speaking. On the other hand, being noticed like that terrified me. As an adult, I still become nervous before big speeches, and frequently find myself wishing I didn't have to appear on stage. There is always a "before-and-after" effect. Before the speech, I want to run away and hide from it. Afterward, I feel wonderful about the attention the speaking brings me. I have even found that I'm a pretty good preacher and have delivered two sermons at my church, usually to good reviews.

I discovered that giving speeches turns out to be a way for an outsider to feel good about himself or herself, without really becoming an insider. When you are on the stage in front of an audience, you are still apart from them, but being an outsider doesn't hurt as much, because you are, in some sense, above them. Really good speakers know how to move people through their talks, and that is truly empowering.

In my mind, even in my 70's, that moment in kindergarten remains a defining experience of my life. The rest of my time on Earth will, I suspect, consistently carry with it the tension between being obscure and comfortable and being prominent and anxious.

However, I have also worked on this issue in therapy and have had some success with it. I have realized that, in my mind, I thought I *had* to get nervous before a speech or public appearance. In other words, there was no way to get the "goodies" without going through hell to get there. I reinforced this concept with stories about entertainers who had stage fright. I saw, for example, an instructive documentary about Elvis Presley before a performance in Las Vegas. He was nervous as a cat, pacing around and talking to his band nonstop, wishing the show would begin, but scared about how it would go.

I was astonished. I mean, here was the King of Rock and Roll, at the height of his career, terrified of performing before his adoring fans! I felt a bit better about my own issues, and I guess I understood why he had started taking so many drugs.

I also read about Carly Simon, and how she had stopped performing for a while because she experiencing paralyzing episodes of stage fright. Again, I felt I was not alone if a star like Carly Simon could have those feelings.

However, in working on this issue, I realized that this stress was not necessary. I came to understand that I do have a gift, whether I am giving a speech, preaching a sermon, or marrying a couple (all of which I have done). God has given me this gift and wants me to use it. I should enjoy it, I should not boast about it, but be proud of this skill, and use it to the benefit of others. Suffering is not a requirement! I do believe, based on experience, that a little nervous energy prior to a speaking opportunity, is a good thing if it is channeled and used well. I've had a few occasions when I have been really relaxed, and my talk fell flat.

At the same time, I have never quite gotten over feeling like an outsider, either in kindergarten or in my neighborhood. There were other ways in which I felt like an outsider. For example, one of the big things to do in the neighborhood was for all the kids to get together and eat watermelon on hot summer days in someone's backyard. Lots of fun and all, but there was one problem: I didn't (and still don't) like watermelon, so my mother would bring along a cantaloupe for me to eat. Of course, that would set off howls of derision from everyone else and I would wish I wasn't there. My mom didn't know it, but it would have been better for me to have skipped these things rather than going there and eating something no one else liked.

Lessons learned:

Later in life, I felt that this feeling of separation helped explain my interest in other kinds of outsiders, ranging from angels to vampires to extraterrestrials. I have learned how to behave like an insider, which is very helpful if one wants to function in this world. However, I have not yet worked through the mentality of being outside the mainstream and I don't know if I ever will. It may well be that my purpose cannot be fulfilled by being on the inside.

Consider this: I am a Christian, and every Christian, by definition, holds outsiders in the greatest reverence. Jesus certainly could have been on the inside. All the evidence suggests that he was very bright, knew the scriptures, and was called "Rabbi" by more than one person. He chose a different path, however, and paid for it with his life. Later, Paul followed in his master's footsteps. He was an honored member of the Jewish faith, and a zealous persecutor of the Christian heretics. He was even a Roman citizen, no small honor for a man of his background. But he had a vision, gave it all up (called all the honors "rubbish")(1) and endured incredible hardships to preach the gospel to the world. He also died as a result of his commitment to his faith.

In fact, all the major religions tend to honor someone who could have been an ordinary person living an ordinary life who felt compelled to do something really quite different.

I do know that some of my most enjoyable memories are those when I felt comfortable and part of a community. This may be the reason I have enjoyed team sports so much. For reasons that are made clear later in this book, I didn't really get into football until I arrived at Harvard and played for my "House" for two seasons. And except for one season, when I "retired" from baseball, I have always loved playing baseball and softball. I still play softball in the Harvard league, and I am almost always the oldest player on the field.

When you are on a team, and the competition gets into your blood, you lose the sense of self, especially of the isolated and socially awkward self. You become one with the game, if only for a few hours. You realize that you are part of a larger whole, and that whole system can only succeed with your assistance. At the same time, while there may be stars on some teams, even they cannot win the day all alone.

To me, this is a perfect experience. It is the balance of unity and diversity that is the great insight of the Overview Effect: you do not need to subsume your individuality to achieve unity. In fact, the more you express it, the better everything will be. (2)

I also felt this way during my (very) brief career in the military, especially at the euphemistically named "Summer Camp" that I attended prior to getting my commission.

This is why, though most people see me as a nerd, bookworm, or something else that has nothing to do with war or sports, I love team sports and the idea of military service. For a long time, I have suppressed those parts of myself, the warrior part, I suppose it is, but now I am going to let them run free!

CHAPTER SIX
NORMALCY

Other than kindergarten, I was living a normal life in Greenwood, Mississippi after we moved into our new house. I remember a lot about those years that is not all that profound, but does portray what life was like back then. I still felt like a loner much of the time, but less so when I was in my neighborhood than when I was at kindergarten.

There were a quite a few kids in the neighborhood and we played together spontaneously, especially on hot summer days. Looking back on that time, it is worth recalling that we did not have television, computers, video games, cell phones, or any of those electronic entertainment media. I am not one to condemn these new technologies. In fact, I like them. However, lacking them made for a lot more interaction with the other people who lived around us, and a lot more inventiveness in our play.

There was a house with an especially large yard across the street from ours, and that was the gathering place. The moms would sit in reclining chairs along the periphery while the kids ran around, barefoot and shirtless on the grass.

As I recall, the house across the street was owned by a family with a boy about my age named Jack Frost (no kidding!). My mother told me he was a cousin, but I am not sure how that was the case, and he and I were never close.

Nobody wanted to wear shoes, but you paid a price because the yard was full of clover, and if you stepped on a honeybee, you got hurt. Even though we didn't have a lot of ways to entertain ourselves, I don't remember it as a time of deprivation. It was an enjoyable period for me.

Up the street a bit from Jack's house lived the neighborhood bully, Charley. He was fat, crude, and aggressive with the rest of us. I was kind of scared of him, but not very much. He was big, but not muscular. I think I knew early on that he was mostly bluster.

One of the big issues in those days was which branch of the service was better. For most of us, our fathers had just come home from World War II, and we argued endlessly about whether the Army, Navy, Air Force, or Marines were best. Charlie's dad had been in the Navy, and mine in the Army, so we were natural enemies. Actually, most of us enjoyed the arguments but did not take them too seriously, He did, and threatened to beat us up if we didn't agree with him.

One day he was over at my house and we were doing something in my bedroom. I can't remember the sentence he spoke, but he ended it with "son," as in "That's not the way it is, son." I hauled off and hit him in the nose without even thinking about it.

"Ow," said the bully, "why'd you do that?" and his nose began to bleed. "I'm not your son!" I said to him.

It was very important to me that he understood that I was my father's son, not his, and I couldn't imagine why he had said such a thing. I was too young to understand it was just a figure of speech.

Of course, he didn't hit me back. Like most bullies, he was a coward when he was confronted. And actually, I was in the wrong, because calling somebody "son" in that way was very common in the South in those days. He didn't mean anything by it. He started yowling when he saw the blood and my mother came running. While she stemmed the nosebleed, she scolded me for hitting a "guest." I apologized, but secretly felt good about it. I had popped the neighborhood bully and he hadn't done a thing about it, except cry.

Charles is not the only person I remember from those days. There were quite a few others who hung out with our little pack, but two stand out, Jack and "Goddy Ralph."

Jack was an older boy, maybe 12 or so, who essentially ran the neighborhood. All of us looked up to him, and he was a benevolent dictator. Unlike Charlie, who tried to rule by brute force, Jack had charisma. He lived in a big house behind ours, and I always assumed his parents were very wealthy.

One day, the neighborhood was abuzz with the story of something that had happened in that house, and it reinforced our view of Jack as a hero. Apparently, a burglar thought the family was rich as well. He broke in late at night and started looking for loot. Jack came running out of his bedroom, jumped on the guy's back and started stabbing him with a kitchen knife. As the story went, the criminal was actually happy when the cops showed up and begged them to take him into custody, anything to get away from this knife-wielding kid!

This just added to the Jack mythos and he really became somewhat larger than life. He was usually pretty nice to us younger kids, but he could be cruel. There was one time when we were playing a game of some kind, and I forgot what I was supposed to do, and he scolded me in front of the others. It really hurt, I must say, and that is probably why I still remember it so vividly.

We believed in Jack's capabilities so completely that when I heard one day that he was building an airplane, I immediately had visions of soaring into the skies and flying to exotic, faraway lands. I told my parents and everyone who would listen that "Jack's building an airplane!"

Imagine my disappointment when I got to Jack's yard and saw some old packing crates nailed together, a propeller he had gotten somewhere, and wheels, origin also unknown. I will admit it did look like a plane, but it was clear, even to me, that this was not a plane that would ever lift off the ground!

I think that is the only time I was ever disappointed in Jack. I seem to recall telling my mother that "it wasn't a real plane," and her gently explaining to me that it wasn't Jack's fault. I had just let my imagination run away with me. Okay, that made sense, but I was still disappointed.

There is one other memory I have of Jack. The day came when we were playing in the backyard of Goddy Ralph (more on him later). There was a huge sandpile there and we were all building things. About all I could manage was a deep hole, but Jack built a beautiful sandcastle. I looked at it, and the sin of envy took over my soul. I wanted it so bad that I asked him to give it to me. He said no, of course, and I burst into tears.

I ran home, crying to my mother, and said, "Mama, Jack won't give me his sandcastle." For some strange reason, I thought she would side with me! To her credit, she didn't. "Honey," she said, "Jack built that castle. There's no reason why he should give it to you. Now, you go out there and build your own castle, or just be happy with whatever you can do."

I was embarrassed because, even at the age of four, I knew my behavior had been ridiculous. I realized that, while I might envy what Jack could do, it was wrong to try to get the product of his work away from him, and very wrong to let people know how jealous I was.

"Goddy Ralph" was a different person altogether. He was perhaps even more outside the mainstream than I was. Today, we would say he was "developmentally disabled." He had some speech problems, and said things like, "Get in cah, go vroom, vroom," when he wanted his parents to take him somewhere. As I recall, he was the butt of some good natured kidding, but I don't think we were as cruel as we might have been. Cousin Anne named him "Goddy Ralph" for reasons only she could understand, but not articulate. So I called him that, and we used to say, "Let's get in the cah and go vroom, vroom," and chuckle. But we never made fun of Ralph to his face. He was a nice kid and deserved to be treated well.

Later in life, I heard he became a famous artist. Like many of the people in my history who were laughed at, he got the last laugh, and I am happy for him.

Those are some of the guys I remember back in the Greenwood days: Jack the Leader, Charlie the Bully, Jack Frost, and Goddy Ralph. I have heard it said that as you age, you have trouble remembering the names of people you have just met, but can recall the names of people you knew half-a-century ago. Well, it's true!

I don't remember any girls being in our neighborhood in my pre Kindergarten days, though I am sure they were there. Being a boy, I guess I just gravitated to the guys and we ran together in "Jack's Gang."

I do remember visits by my cousin Anne (now known as Harriet). We got along well, and enjoyed being with one another, but we fought a lot. I remember not understanding why that happened. It was a new kind of conflict for me, and different from what I experienced with the pack of boys. I knew, somehow, that Anne and I were close, like brother and sister, so it seemed wrong to me that we would fight. I also knew I shouldn't hit her, like I had hit Charlie. It was clear to me I did things that frustrated her, and I wasn't sure why.

However, I was told that brothers and sisters did fight. I didn't know anything about that, since I was an only child.

I never asked my parents why they didn't have more children, and I wish I had. Being an only child has shaped my life in significant ways, yet I don't know whether Mom and Dad simply couldn't have any more kids, or chose not to, which seems unlikely. I say this because of something that happened when we moved to Germany, but I will save that for later.

I have one other memory of those days, which is the time when I saw "witches" in the neighborhood. I was in the side yard, looking toward Grand Boulevard, which was the "main drag" not far from our home. A car pulled up in front of a house some distance away and across the street from "Goddy Ralph's" place. Two women in black robes got out.

They went into the house, and I feared for the owners. As far as I knew, they were about to be cursed, or even worse, about to die. I ran into my house, calling out for my mom, "Mama, mama, there are witches in the house across the street!"

She came out with me, and the "witches" were leaving around the time she did. She took one look, laughed, and said, "Those aren't witches, they're nuns."

Of course, I had no idea what a nun was, or why they would dress like witches. My mother explained it to me, and I felt much better. It reminded me of the day I told my mother I was dying, and she laughed and said, "No, you have the flu."

CHAPTER SEVEN
FOOTBALL AND BASEBALL MEMORIES

My love of sports began at an early age in Greenwood. I remember going to my first Greenwood High football game at the age of four or five. I loved the color and the noise of the bands and the crowd. I became somewhat famous with my parents' friends because I could recall the scores of all the games. Today, I only remember that Greenwood beat Corinth 60-6 in that first game I attended.

Anyway, I enjoyed football and was really into the Greenwood team. There was one particular player that was the hero of the team, Jimmy Lear. He was the halfback, he ran like a deer, and scored touchdown after touchdown. To say that I idolized him would certainly be an understatement.

My father was co-owner of an engineering firm he had started after getting out of the Army, and he had high school students work for him during the summer. One day, he came home and told me something that electrified me: Jimmy Lear was working for him, and would I like to meet him?

At that moment, I was more impressed with my dad than ever, and really excited beyond words. "Would I like to meet Jimmy Lear? Would I?" I couldn't wait for the day that Dad brought

Jimmy over during their lunch break. I remember getting dressed in clothes I thought were appropriate to the occasion and being ready far before noon. I fidgeted and thought about what I would say. It was very similar to how I prepare for a big meeting as an adult. Am I wearing the right clothes? How will the conversation go?

I don't have a good memory of him, of course, but I think of him as being wiry, muscular, and blond. Not a big guy, but the kind of small player who could easily outrun enemy linemen for a touchdown, which I had seen him do many times.

I also don't remember what we discussed, in any detail. I know he asked me if I wanted to play football, and I said, "yes" enthusiastically. He also asked me which school (college) I wanted to attend, and I said, "Auburn." I had already prepared that answer, since I assumed he would ask the question. However, I knew nothing about Auburn except that I liked the sound of the name.

He said, "Oh, that's a good school."

I don't recall anything else from that meeting except that I felt quite a sense of accomplishment that I had met my hero. At the same time, I had a feeling that has recurred many time as an adult, which is that my larger-than life idol was just another human being when seen in person. If you have ever had a chance to watch actors on a movie set, for example, you are likely to say, "Is that really him, or her?"

When you are watching a halfback run down the field from high up in the stands, they are wearing the bright colors of your school, the bands are playing, and the crowds are cheering, they take on an aura they cannot have when you meet them in the flesh.

There are a very few people who are charismatic regardless of the setting. Bill Clinton is one of them. He is just as impressive in person as when you see him on television. However, this is not the norm, and I suppose I learned it at an early age when I met Jimmy Lear.

Still, I was pleased after the meeting with him, and I really appreciated that my father had been so thoughtful as to arrange it.

Anyway, I retained my interest in football, and then got really involved with baseball as well, a love that has manifested itself in actually playing the game (well, softball) to this day.

I became a Brooklyn Dodger fan, and the way it happened is kind of interesting. For some reason, I hated the Dodgers for a long time. I think it was because they were in New York, and I thought of New York as all that was evil about the North.

However, I learned that the Greenwood Class C minor league team was a farm team of the Dodgers and I also started going to their games. That was it, and I became a Dodger fan! And, of course, I also transformed into a hater of the New York Yankees! So I started going to the Greenwood games. I remember a simple stadium with wooden seats and games played under the lights. If you have ever watched the movie "Bull Durham," it was something like that. There was a third basemen with white-blonde hair and I think his name was Devlin. We joked that he was the "devil" because he was always up to some mischief.

There was a shortstop who was compact and slim with black hair named Dolan. He was my hero, the Jimmy Lear of my baseball days. Devlin and Dolan, they were my favorites.

When my mother or father would agree to take me to the games, I was in heaven. I recall that there was an old man who often attended, and he enjoyed sitting with my mother, my friends, and me. He would tell us kids to shout out, "Kill the umpire!" and we would enthusiastically comply. My mother would say, "Frank, don't say that!"

I would look at her and say, "Mr. Jones told us to."

Then she would say, "Mr. Jones, don't tell them things like that." He would just chuckle and stop for a while, then later on, he would tell us, "Say 'kill the umpire,'" and we would, and eventually my mother would sigh and give up. I doubt that the umpire could hear us, anyway! This was one of the first times that I saw a crack in the front that adults usually put up when telling kids what to do. It puzzled me, but I liked it, and I loved yelling, "Kill the umpire." I also liked the fact that my mom gave up and didn't push the issue.

I remained a Dodger fan until they moved to Los Angeles. I could not relate to them any more when they made the move. Of course, I had a ready made team waiting in the wings, i.e., the Boston Red Sox. My father had been a minor league ballplayer in their farm chain for about a year and a half, so I decided they would be my team. When I came to Boston to attend Harvard in 1962, it became even easier. And, of course, the Sox and Dodgers both had an implacable hatred for the Yankees.

Anyway, as I look back at my first six orbits as a crewmember of Spaceship Earth, I realize how big a part sports played in my life, and I feel good about it. Baseball and football introduced me to the joy of playing on a team. Many years later, I love playing softball on the Diamond Dogs, a part of the Harvard staff members' league. I am almost always the oldest person on the field!

CHAPTER EIGHT
FIRST GRADE

In comparison with kindergarten, I have very few memories of first grade in Greenwood. I recall that we walked a relatively long way to school, including crossing Grand Boulevard. This is something, sadly, that might not happen today, when there is so much fear of abductions.

I also remember a typically small schoolhouse set in a green park-like area. I liked school as compared with kindergarten because it was the closest thing to being intellectually stimulating that I had experienced up to that time.

One small regression about my intellectual development at that time, such as it was: my mother used to read from picture books to me every afternoon before I took my nap. I would watch and listen as she went along, and I was deeply interested in this whole process called "reading." I recall that my favorite book had a group of animals in it that were always going off on great adventures. One day, as Mom was reading and I was following along, I had an epiphany! *I could read the words myself!* What a moment of great power that was.

The ability to read before I entered first grade gave me an extraordinary advantage over the other kids, who generally could not yet read, and that was in fact their first task on getting to school.

That was where they learned how to read, but I already had mastered the skill. Looking back on it, I think that is the source of an overall feeling of, once again, being an outsider. People thought I was *a lot* smarter than everyone else, but I'm not sure that was true.

As I think about it now, being able to read when my peers could not was like the special powers of a superhero, x-ray vision, or the ability to fly. It was effortless to be better than others in that situation. Of course, I also won all the spelling bees because I had a large and growing vocabulary at an early age.

This early edge has carried me forward throughout my life. Combined with my experience in public speaking, it has given me the power to move people through my words in both writing and speeches. I still read voraciously and retain all that I read pretty effectively. Everyone keeps telling me to go on *Jeopardy* because I do very well at that game when watching it at home on TV (I have actually entered the online competition to be selected but have heard nothing back.)

The truth is that I am *not* more intelligent than other people when it comes to subjects like math and science. It is only in subjects where my early reading skills give me an advantage that I can stand out. This superiority that I had at an early age had both benefits and drawbacks. The teachers loved me; but the students did not. This is understandable I made the teachers look good, while I made the students look bad.

There is a saying I have heard about why bullying is a major problem in Japan. It is said that "the nail that stands up soon will be hammered down."

That is how I felt growing up. To avoid taunting or worse from my peers, I tended to downplay or even hide my skills.

I have rarely felt this superiority in other areas. Generally, I have to compete like anyone else to excel at anything. The closest analogy I have found may sound a bit odd. However, it lies in the realm of ping-pong. I have recently found that I can beat all of my sons-in-law at this game. In fact, I have never lost to any of them. At one point, when I was playing one of them I felt an exhilaration that was out of the ordinary, because I simply knew that I would win, and I wished I could extend that feeling to the other domains that had greater meaning to me and my family, like earning money.

Anyway, this advantage served me well in first grade, but also had the negative consequences I have already cited.

I don't remember many of the other students at that time. There is one I recall, because she had an odd name. I won't repeat it here to avoid embarrassment, so let's just call her Tracy. Even at an early age, I knew that she had an unfortunate name, and that she did not deserve the teasing it generated. I also recognized that she was a nice person, and someone who was really smart, but that all of these virtues would not help her to overcome that name.

Interestingly, while I was feeling all this compassion for Tracy, she is the only person I can remember who made fun of me in first grade. We were working in our coloring books and the subject was cardinals (birds, not religious leaders). The male cardinal was easy, of course: red, with black above the beak. The female was another issue altogether. In pondering what the female coloring might be, I concluded that it was not quite brown or green, and I experimented by combining different colors together. I used red and green, as I recall.

Anyway, Tracy saw what I was attempting and said, "Frank, that's really dumb. The females are brown. What are you doing?" I was more surprised than angry or hurt when she said that. First,

I didn't agree with her about the color, and second, I was innovating, couldn't she see that? This is a feeling I have often had in my life. People often find what I am doing to be wrongheaded or beyond their understanding. Innovation and experimentation don't interest them, but it fascinates me.

Anyway, since that moment at the age of six, I have had to decide how to handle that kind of joking about my interest in exploring paths that others do not want to take. It still bothers me when people make fun of my ideas, but I'm used to it now.

Otherwise, first grade was uneventful, but that didn't last for long.

CHAPTER NINE
CHURCH

My mother's family were Presbyterians and my father's family were Methodists. I would say that on my mother's side, people were "religious." By that, I don't mean that they made a big deal of their faith, but they were faithful and went to church every Sunday. My mother also thought it was important for me to go to Sunday School. More on that later.

My father did not go to church in those days, though he later did. A man of few words, the only thing I ever heard him say about religion focused on what he observed with the indigenous people and the missionaries in New Guinea during WWII.

"The missionaries tried to convert the natives," he said, "and I thought it was foolish because those people were perfectly happy until the missionaries came along and started preaching to them."

Anyway, my mother and grandmother took a different approach to religion: they thought it was important and that I was going to get a full dose of it. I was actually kind of like the people who lived in New Guinea. I was perfectly happy until the preaching started.

The whole thing was kind of confusing to me. My mother gave

me a picture book about the life of Jesus after I learned to read. I was absorbing it on a fairly long car trip, sitting in the back seat.

As I read about Jesus, I became increasingly excited. Wow, here was someone I could really respect and emulate. He cured sick people, he was very kind, he had really good ideas, and he didn't bully anybody. What a guy!

When I got to the end, I was shocked. They killed him! Why did they do that? He didn't deserve to die, all he did was good. I asked my mother to explain it, and she said, "That's why I wanted you to read it, so you could think about that."

I was still puzzled, and I guess I am to this day. However, having experienced quite a few betrayals myself, I am beginning to understand it better. Just because you are a good person doesn't mean people are going to be nice to you in the end.

Reading that book began a journey that continues for me: defining my relationship to Jesus, Christianity, and the church itself.

Growing up in Greenwood, I had no choice but to be religious in the sense that I went to church every Sunday with my mother and grandmother. I remember that my mom talked to the minister about the fact that I could already read, and I would come to church if I could read my comic books while the service went on.

I find it remarkable how much I remember about the minister and the dialogue between him and my mother. His name was Van Arnold, and he was a really nice guy. He liked to restore antique cars, and I recall going to his house once and seeing a shiny old automobile in his garage.

He told my mother, "I would rather have Frank in church reading comic books than not being in church at all."

One reason I remember this conversation is that my mother

thought it was funny and told people about it every chance she could get. I think she saw it as a victory, in that she had gotten me to do what she wanted me to do, with little or no struggle, and the solution had been blessed by the minister. Of course, it was also linked with the fact that she had inadvertently taught me to read at an early age.

It was interesting to me because it was one of the first times I saw an adult make a thoughtful decision and compromise, rather than just going by the rules and making me follow them.

Anyway, I have vivid memories of reading my comic books and half listening to Reverend Arnold and the choir singing. It was a kind of multitasking at an early age, I suppose, something I have continued into adulthood. From his point of view, I think the Reverend Arnold was right: I got into the habit of going to church, and associated it with something I liked to do, because I also had the memory of reading the comic books.

Now, many years later, I am quite active in my church, and I don't know if that would be true if the idea of being there on Sundays had not been inculcated in my mind at an early age. Moreover, the policy of letting the children come and read their comic books or color in their coloring books is followed in my church as well!

Church in Greenwood was okay but Sunday School was another thing altogether. I did not like going there, and I made up every excuse I could imagine to avoid it. What was the problem? I don't know, but I suspect it was interacting with the other kids that I did not enjoy.

When we went to church, I could escape into my own world of reading, even though there was a lot going on around me. Sunday School required a lot more involvement with other people, both adults and kids. Also, I think I preferred to learn the religion on my own, as I had done with the book my mother gave me about the life of Jesus.

Whatever the reason, the Sunday School issue was a matter of some contention between my mother and me. I came up with every excuse I could imagine to get out of it, usually that I was sick, and she devised every stratagem she could create to get me there.

It was not pleasant for either of us.

Still, I made it to my share of Sunday School classes, and learned a bit along the way.

Anyway, I'm grateful to my mother and grandmother for starting my religious education at an early age. Though I wandered far afield from my original faith, I have come back to it once again. I'm going to church on Sunday mornings, and I no longer need the comic books—and I now see young children doing exactly the same thing that I did, so long ago!

CHAPTER TEN
THE PLANTATION

There is one set of memories that should not be ignored: Mayday. That was the name of the plantation on the outskirts of Greenwood that my Great-Uncle Matt owned.

Uncle Matt was a throwback to the early days of the South. As I remember him, he was big, gruff, drank a lot, and lorded it over everyone who visited his kingdom. I cannot say that I liked him. I wasn't afraid of him, really, but he was an authoritarian and he intimidated me.

I was afraid of Mayday. I understood my little neighborhood in Greenwood, its boundaries and dangers, such as Charlie the Bully. But the plantation seemed endless and mysterious, and there seemed to be a frightening experience around every corner.

It would all begin when we drove down the driveway toward the big house. Polly the Parrott, who lived in a huge cage near the drive, always cried out loudly when we entered his turf. I think he said, "Polly wants a cracker," but I might be making that up. We were told to stay away from him because he would bite your finger with his sharp beak. I didn't have to be told twice.

Then, there was Julia, the huge Great Dane that looked like a horse to me. Julia was taller than I was, and she could look right in my eyes, slobbering from her enormous mouth. She enjoyed chasing me around the house while the adults sat on the porch, laughing. I remember running around the house in sheer panic, screaming to my parents to help me and being really angry at their laughter.

Now, I realize that Julia was just having fun and the family was laughing because they knew I wasn't in danger. But nobody told me that! There was one other animal I remember well, and that was Tony the Pony. He was a benign presence in my experience of the plantation, but he was used by others to create new terrors. Tony was a black-and-white Shetland pony, and he was nice and docile. However, we never rode him alone. Instead, we were accompanied by the African American kids, sons of the sharecroppers who lived on Uncle Matt's land. That's where the trouble began. Let me step back a moment and talk about our relationship with these kids.

I admired them, to be honest. They were amazing in how comfortable they were in their environment, harsh as it was. I want to be careful here, with some 65 years of perspective on the situation. To be honest, the African Americans on the plantation were only a few steps away from slavery. In fact, other than the legalities of it, I don't think there was a lot of difference, and that is not all right with me today.

At the age of four or five, though, I didn't think much about the big picture. What I thought about was that I almost never interacted with the African American population of Greenwood, with some exceptions I will detail later.

I don't recall thinking these kids were inferior to me, or that they ought to pay me any respect (they didn't!). To me, they were just kind of *different*, and really scary. When I went out riding on Tony with them, there was no one else around, and they could do pretty much anything they wanted to do with me.

In most cases, they were older, say 8 or 9, and I had no defense against them.

They rode mules bareback and barefoot, and they seemed to have an almost intuitive ability to control these stubborn animals.

Anyway, back to Tony. I had no idea how to saddle him, so my guides did that for me. That worked out fine except one day, Tony decided he had had enough and began to run away with me, the mule riders in hot pursuit. I was scared and shouting, "Whoa, whoa, whoa!" But he kept going, until the saddle began slipping out from under me and between his legs. That was a bit disconcerting because I was pretty sure I was going to fall off. However, I held on, and he couldn't run any more with it down under his belly.

The two black kids rode up, just laughing their heads off. I was angry with them for laughing at me, but what could I do? Since then, I have connected with three cousins who also spent time at Mayday, and they remember Tony being an ornery fellow who was difficult to ride as well, so I suspect they laughed at other riders as well.

Now that I have some distance from those days, I kind of hope it brought some pleasure to their lives, which were pretty deprived, and I feel guilty about that.

Anyway, our interaction with the black kids didn't end with pony rides. My cousins and friends and I played baseball with them, and tackle football.

I have vivid memories of those games, which were also scary, but a lot more fun than the pony rides. They were played on a wide expanse of lawn near the big house, and the adults were usually close by, so all of us had to behave, to some extent.

We also played tackle football and that was really rough. The plantation kids were a lot tougher than we were, but we held our own against them. At one point, my mother told Uncle Matt that she was worried that someone was going to get hurt. He dismissed her fears with a wave of his hand. "They'll be all right, Mary Ann," he said.

I was so afraid of so many things on the plantation, it's surprising I wasn't afraid of those football games.

Language was a problem at times. I remember one kid saying to me, "Whea d'you stayat?"

"What?"

"Whea d'you stayat?"

"Sorry, what?"

Finally, somebody intervened. "He wants to know where you live (Stay at.)"

"Oh, Greenwood."

When I think about growing up in Mississippi, one of my most significant insights is that racism really is learned. Throughout this book, I will be sharing how I interacted with African Americans in a segregated society, and later in an integrated setting, and I am pretty convinced of this: I wasn't a racist as a child, but I absorbed the thinking of white culture as I grew up. It's been hard overcoming that mindset, and it is a daily struggle, but it can be done. I remember Dick Gregory once saying something like, "I don't care what you think of me, but I do care how you act toward me."

Of all the things that happened on Mayday that frightened me, one stood out: Ed Hunter and "Little Matt" on the swing. They had this homemade swing that was really just a wooden plank tied to some rope and strung up in the limb of a tree that stood on a bluff over the Yazoo River.

They would get on the swing, get it going as high as possible and then leap out into the river and swim back to shore. Whenever

I watched them doing this, they would always say, "Hey, Frank, wanna try it?"

I couldn't swim, and the idea terrified me. They also threatened to simply throw me in, a statement that I took seriously. I still have a vivid memory of Ed Hunter drawling, "Doncha wanna give it a try, Frank?" while I watched Little Matt swing far out over (what seemed to me) the raging waters of the river.

"N-n-n-no," I said, my voice quavering. I was pretty sure they would throw me in and I would drown if I didn't keep a close eye on them. Many years later, I saw Matt and his wife Val in Miami. I told him this story, and he said, "I'm sorry for almost drownin' ya." I told him that he hadn't almost drowned me, just scared me because I thought he was going to! I have other fleeting memories of the plantation. One day, Uncle Matt took me out with him for a ride around the place. I recall seeing a chicken coop with lots of chickens running around in it, and Uncle Matt talking to the people who worked there. I also recall that he put me on his lap and let me drive the car. That was almost a disaster, because we came very close to going into a ditch!

I remember a massive dinner by candlelight in the huge dining room. Black servants were everywhere, bringing in dishes for the diners, as Great Uncle Matt held court at the head of the table. Looking back on that memory, I think the scene must have been very similar to antebellum days. In fact, it was only about 80 years removed from those times.

One of my favorite things to do, when my parents would allow it, was to get away from everyone and go to the library, which was tucked away in a quiet part of the house. There, I would read exotic books like *Bomba, the Jungle Boy*. When I was in the library, I felt safe, and also inspired by what I was reading. I didn't have to worry about runaway ponies, big Great Danes, tough black kids, or anything else. I knew I was being anti-social and that it was a kind of guilty pleasure, but one of my fondest memories was of being in that library on a rainy afternoon and reading *Bomba*.

[This feeling of safety has persisted over the years. By analogy, my freshman year at Harvard was marked by studying in Lamont Library almost every night, always in the same spot in the main reading room. I enjoyed it, even though I was technically not alone. I was with many other freshmen, working hard and trying to meet the standards of one of the world's great universities. However, it was quiet, and back in those days, you were allowed to smoke, so I would stoke up my pipe (does anyone smoke a pipe anymore?) and study for three or four hours, going back to my dorm when Lamont closed.]

Overall, when I look back on the plantation days, they were both frightening and exhilarating. Generally, when I have recalled those times, I've remembered the situations that frightened me, like Tony the Pony running away with me. However, there were other instances when I was not afraid and I could have been. So those visits to Mayday were not all good and not all bad. Yes, I was scared at times, but I was also pretty brave at other times, and that's the lesson to be learned from those days.

CHAPTER ELEVEN
MY PARENTS

I have, of course, spoken of my parents on several occasions throughout this portion of the book. Inevitably, they appear again and again as I tell the story of having my tonsils out and the time I thought I was dying but only had the flu.

However, there are some other recollections of them that are worth sharing.

For example, I have realized while writing this memoir, that my father saved my life one day. Billy and Betty (my Aunt and Mom's sister) were visiting us in Greenwood, along with Anne and Leigh. At the time, I guess Leigh was about two and Anne and I were around five years old, something like that.

Anyway, our parents gave us some Lemon Drops and told us not to let Leigh know we had them. So we were cavorting around the house, laughing, but keeping our mouths tightly shut so that Leigh wouldn't know we had the Lemon Drops.

Suddenly, I swallowed the candy whole and began choking. My father picked me up by my ankles, held me upside down, and pounded on my back (there was no Heimlich maneuver in those days). The Lemon Drop came out, I coughed and gagged and after some nervous fluttering around to be sure I was all right, we went ahead with our family day.

This is, of course, an incident seared into my memory because it was so dramatic. However, it was not until I began writing this book that I realized *my father saved my life.*

You might say "Well, that's obvious, of course, and you have had this memory all this time, so what's the big deal?"

The big deal is that I have finally felt some gratitude for what my dad did. If he hadn't acted so quickly and so forcefully, I might have died and I might not be writing this story. Like a lot of young people, my thoughts about my parents have been sullied by conflicts and issues that developed over years of interactions, not all of them very good. That can get in the way of memories like this one, in which I think of what might have happened that day. Maybe someone else would have saved me, maybe the Lemon Drop would have dislodged on its own. I don't know. All I know is that my dad acted fast, and I am here to tell about it!

Once, my father brought home some little puppies that he had found abandoned on the side of the road when he was out working. He put them in a box with a blanket and tried to give them warm milk from a bottle. Mom told him it would never work; they needed their mother. He persisted, though. I don't recall what happened with the puppies, but I suspect my mother was right, and my father wasn't able to save them. Still, I have always admired him for trying.

I also recall that my dad had a darkroom in the basement of the house. He had been into aerial photography all his life, had been in photo intelligence during the war, and had continued his interest in photographic matters after getting out of the Army when World War II ended. I found it fascinating that he had a skill like that, and I appreciated it when he let me into his special retreat, but I didn't like the smells and the process seemed a bit tedious to me.

In my early life, I only remember my father getting angry with me once. There was a social gathering at our house, and it took place in our yard. I don't recall who was there or why, but I got upset for some reason and ran away from the scene, crying. My father left the event and came after me. I recall standing next to him and near some of the guests' cars, and he said to me something like, "I'm not gonna have you goin off and boo-hooin like this when we have guests. There's no reason for it."

Even at my young age if found the term "boo-hooin" to be funny, but I dared not laugh. I don't think I had ever seen Dad get angry like that, so there was no telling what he might do if I laughed *at* him! I guess I remember this incident so vividly because it is one of the only times my father lost his temper with me, and it is the first and last time I heard anyone say "Boo hooin'."

As for my mother, I have already reported most of my memories of her, though there is one other to mention: my ill-fated effort to learn how to swim, which she initiated. Looking back on it, she was right. I was three or four years old, and it was natural that I would learn this essential skill. (remember my cousins, the swing, and the river?)

So Mom took me down to the municipal pool, which was full of screaming kids running around and jumping into the water. I was already freaked out because, as you know by now, that was not my kind of scene.

The instructors had us jump into the pool in the shallow end, which I did. Then they had us put our faces underwater, which I did. I am not clear on what happened next, but at some point I got a lot of water up my nose, felt like I was drowning, and announced to my mother that I wanted to go home right then and there.

I don't want this to sound critical of my mother, because she made the kind of compassionate decision I am always advocating for parents, bosses, and leaders to make. Rather than forcing me to continue, she said, "Okay," to my great relief. So we went home and the swimming lesson was over. I am sure she thought that by not forcing me, I would go back in my own time.

I recall that later in the day, a neighbor asked me how the swimming lesson went and I told her what had happened. I felt really good that my mom had supported me in coming home, and because of that, I wasn't ashamed to let the neighbor know what had taken place.

The problem with my mother's decision to let me get out of learning to swim is that I did not go back and I never learned properly, which kept me from many enjoyable pursuits over the years and, on a couple of occasions (to be discussed later) almost led to my demise. At this point, I have gone from saying that I can't swim very well to saying I really can't swim. Now my wife Donna is trying to teach me, and she is having some success, so the story may have a happy ending.

For now, I will just say that being a poor swimmer contributed to that feeling of being a loner, and not part of various groups, over the years. I can recall many times when friends were doing something that related to water, and I begged off to do something else. In fact, as we go through this little autobiography, you will read many examples of it.

My parents also went out of their way to make me happy in those days. I was quite the fan of Hopalong Cassidy, a cowboy hero of that time. We didn't have television, so I had only seen him in black-and-white movies (maybe in comic books as well). Anyway, he was my favorite, more than Roy Rogers, Gene Autry, Tom Mix, the Lone Ranger, Lash LaRue, and others of that type.

Well, one day, out of the blue, my father informed me that "Hoppy" was going to be in Memphis, and did I want to go see him? Well, of course, you know the answer! There were many exciting elements to this plan. For one thing, Memphis was, for me, a very distant place. I mean, it was even in another state! Actually, it would only be two hours away today, but before the Interstates were built, the trip was much longer.

It also meant that I would be traveling in the car with both my parents for a long time. I did go places with them, of course, but this would probably be our longest trip. I liked traveling with my parents, so this was going to be a great trip.

I don't remember much about traveling from Greenwood to Memphis, probably because I was so excited to see "Hoppy." I do recall getting to a department store and joining an extremely long line that seemed to wind its way up some stairs to a higher floor. I remember being tired and shuffling along until, all of a sudden, there he was! But what a shock!

He was in his black cowboy outfit, with silver buttons, a silver belt buckle, and so on, and he was nice enough when he shook my hand. But he was old! Really old! I shook his hand, mumbled something in response to his cheery "Hello, there," and shuffled along out of the store.

When we got back in the car, my parents asked me how I had liked Hoppy. I was honest: "He was old, as old as you, Daddy!"

They both laughed at this unexpected response. Of course, my father was about 35 years old at the time, so it was a funny comment. In fact, William Boyd, who played Hopalong Cassidy, was 20 years older than my dad, and in his mid-50s when I saw him. So it's no wonder that, without the magic of Hollywood film illusion, I was shocked.

If you look at the William Boyd Wikipedia site, you can see him with a bunch of little boys about my age at about that time.

http://en.wikipedia.org/wiki/William_Boyd_(actor)

Looking at Wikipedia reminds me that William Boyd actually looked a lot like my father, which might be why he edged out Roy Rogers and all the others as my favorite.

Anyway, I recovered and was pretty cheerful on the way home. I guess it might have been the first time I ever had the experience that there is illusion and there is reality, and they are not the same. It was clear to me that something was done in the movies to make Hopalong seem younger. And of course in the comics they could make him any age they wanted.

I suppose he was still my favorite after that, but things would never be quite the same.

Nevertheless, my parents and I had quite a conversation on the way home. I recalled that Hopalong had a horse named Topper who had not, of course, accompanied him to the department store. I told my parents that, now that I had seen Hoppy "in person," I wanted to see Topper "in horse."

They laughed and I realized that I had made something up that was not quite right but was clever. I could tell they were not really laughing at me, but with me, and that was a good feeling.

It was getting dark and there were crimson clouds in the sky that looked like fingers. Somehow, we got to talking about "Taps," the bugle melody that is played at the end of the day on military bases, and also at military funerals. I said that I knew most people felt sad when they heard it, but I didn't. I said something like, "I think it's beautiful, and I am not afraid of death."

This time, my parents didn't laugh, but indicated that they thought I was saying something profound for a little kid.

I think I have retained these memories because this is probably the first time I can remember impressing my parents with my intelligence in quite that way. The thing about seeing Topper "in horse" was funny, but took some thought. The thing about "Taps" was more the sign of deeper thinking and emotion than one might expect from a four-year-old.

I learned another way of getting positive attention that day, i.e., by sharing my thoughts. And I have been doing it ever since!

CHAPTER TWELVE
WHAT I HAVE LEARNED

I have found it to be both enlightening and therapeutic to look back and "look down" on my life during orbits one through six. A lot happened during those years, beyond the narrow perspective that I had growing up in Greenwood, Mississippi.

Most momentous, of course, was the end of World War II. Allied troops fought their way into Germany and brought down the Third Reich. President Roosevelt died, and was replaced by Vice President Harry Truman, who made the decision to drop the atomic bomb on two Japanese cities, Hiroshima and Nagasaki.

Then, there was the postwar phase, when all the surviving service men (and women) came home, like my father. A new war began, the so-called Cold War between the West and the East.

I was, of course, blissfully unaware of all of this. After all, I couldn't read until I was around four years old, and even then, I wasn't reading newspapers yet, just comic books. Certainly, my friends and I were hyper aware that our fathers had been in the war, and were sensitive to which service was the best (as I have discussed earlier). But we lived in a relatively safe little bubble in that Greenwood neighborhood. Given some of the negativity that was slated to intrude into my life later on, I can now see that this started out as a good time, in many ways.

There is one simple, and not very profound, insight that I have gained from this writing, i.e., why I like contemporary, or modern, houses. I have always liked them and wanted to live in one. Well, duh, in writing this document, I have realized that the house in Greenwood that my parents built was exactly that kind of house! Mystery solved!

I can clearly see the traumatic impact of my grandfather's death. That event hovers over the relatively idyllic days like a dark cloud obscuring the sun. In retrospect, I don't think he would have done it if he had realized how much it would hurt the rest of us, including me.

I have also learned that I need to get beyond this trauma because it has affected me in every aspect of my life. Every loss of a friend, job, or any kind of relationship throws me back into the loss of my grandfather, and causes me to feel that I am worthless. Of course, his death was not about me, it was about him. However, when you are three or four years old, everything is focused on you, and you feel that you are the cause of everything. I felt that I was not important enough to him for him to stay around, and to some extent, that was true.

He was in so much pain that he had to end it, even though it would, in a sense, transfer his pain to me. As I look back and down on that time, I now see that I have the choice to let go of that pain. It is not mine. I don't have to carry it around anymore. I don't need to feel abandoned any more, and I don't have to re-experience it when I lose something, or someone, in the present.

This one insight is worth all the writing I have done in the first section of this book. Even though I have studied, practiced, and experienced quite a lot of therapy and personal growth over the past 30 years or so, this is quite a revelation. I thought I had worked through all of my issues with my grandfather, but it is clear that I had not.

Just from this writing, which is occurring in July of 2013 (with a later edit in July of 2016), I am seeing clearly how an event in 1947 continued to shape my life, even 66 years later. I find this amazing, but feel strongly that it is true.

Another event that stands out is that my father saved my life. This may seem overly dramatic, but until writing this book, I had not seen it that way. Having that perspective wipes out a lot of the anger I felt for my dad in later years. I can see his deep love for me in retrospect, not only in that act but also in countless others. I think we have such a strong need for our parents to be perfect that we become irrationally angry with them when they have flaws.

I didn't really overcome my anger with my father until I was in my 30's, and it wasn't because I remembered that incident. It happened through a process that I will describe later in this book. Anyway, the older I get, the more I see what I owe to him, and how much I am like him. And recalling this incident is an important part of the healing I have been experiencing.

So there is a lot of forgiveness in this remembering. I am forgiving my grandfather for leaving me, and my father for the flaws that developed in him later in life. I am forgiving my mother for not making me learn to swim, and lauding her for teaching me to read. It is only through this writing that I have seen what an amazing gift early reading was, and how it gave me an advantage that has carried through my entire life. In fact, it has truly defined my career, which is that of a writer.

Overall, I am far more grateful for my life than when I began this project and that has made the entire process more than worthwhile.

Made in the USA
Monee, IL
06 November 2026

68012411R00042